Crawly Creatures

MILLIPEDES

GAIL RADLEY

MW01517591

BLACK
RABBIT
BOOKS

Bolt is published by Black Rabbit Books
P.O. Box 3263, Mankato, Minnesota, 56002.
www.blackrabbitbooks.com
Copyright © 2020 Black Rabbit Books

Marysa Storm, editor; Grant Gould, designer;
Omay Ayres, photo researcher

All rights reserved. No part of this book may be
reproduced, stored in a retrieval system or transmitted in any form
or by any means, electronic, mechanical, photocopying, recording, or
otherwise, without written permission from the publisher.

Names: Radley, Gail, author.
Title: Millipedes / by Gail Radley.
Description: Mankato, Minnesota : Black Rabbit Books, [2020] | Series:
Bolt. Crawly creatures | Audience: Age 9-12. | Audience: Grade 4 to 6. |
Includes bibliographical references and index.
Identifiers: LCCN 2018018968 (print) | LCCN 2018021580 (ebook) |
ISBN 9781680728170 (e-book) | ISBN 9781680728118 (library binding) |
ISBN 9781644660225 (paperback)
Subjects: LCSH: Millipedes–Juvenile literature.
Classification: LCC QL449.6 (ebook) | LCC QL449.6 .R33 2020 (print) |
DDC 595.6/6–dc23
LC record available at https://lccn.loc.gov/2018018968

Printed in the United States. 1/19

Image Credits

Alamy: Daniel Borzynski, 26 (top);
MichaelGrantWildlife, 11; bigking.info: LianTze
Lim, 23; bugguide.net: sfloydevans, 18; commons.
wikimedia.org: Bjørn Christian Tørrissen, 20–21; Franco
Folini, 17 (btm); Jörg Spelda, 16 (btm); Marek, P.; Shear, W.;
Bond, J., 28–29; Sarefo, 16 (top); Stemonitis, 17 (top); Dreamstime:
Kriangkrai Saikasem, 17 (middle); Woravit Vijitpanya, 14 (middle);
orkin.com: Orkin, 22; pestwiki.com: Unknown, 22–23 (top); pickle.nine.
com/au: Pickle, 29; Science Source: Fletcher & Baylis, 4–5; Nicolas Reusens,
6–7; Shutterstock: Adisak Rungjaruchai, 8–9; clarst5, 25 (bird); crystaltmc,
14 (top); D. Kucharski K. Kucharska, 28; docter_k, 14 (btm); enterphoto,
31; Eric Isselee, 25 (mouse); Federico.Crovetto, 25 (lizard); Iuliia Azarova,
25 (mushrooms); kamnuan, 25 (millipede); Kritsana Pitlertpitak, 22–23
(btm); Martial Red, 12–13 (silhouette); MR.Phakpoom Mahawat, 1; Nin-
ey Azman, 19; oumjeab, 25 (dead plants); PK.Phuket studio, 25 (living
plants); Sakdinon Kadchiangsaen, 12–13 (bkgd); Stockcrafterpro,
Cover; Yogesh_more, 3; Yuttana Joe, 32; simple.wikipedia.org:
Esculapio, 26 (btm)
Every effort has been made to contact copyright holders for
material reproduced in this book. Any omissions will
be rectified in subsequent printings if notice is
given to the publisher.

CONTENTS

CHAPTER 1

Meet the MILLIPEDE

Night falls in the forest. Hundreds of millipedes crawl out of hiding. With wavelike movements, they creep along the forest floor. Moving slowly, they wander through the damp dirt and leaves.

COMPARING SIZES

bristly millipede

greenhouse millipede

American giant millipede

giant African millipede

inches

A Lot of Legs

Millipedes are **arthropods** with smooth, rounded bodies. Their bodies have many parts. Most parts have two sets of legs. There are more than 10,000 known types of millipedes. They come in many different sizes. Most types have between 40 and 400 legs.

.07 to .12 inch (.2 to .3 centimeter)

about 1 inch (3 cm)

up to 4 inches (10 cm)

up to 12 inches (30 cm)

2   4   6   8   10   12

SEGMENTS

LEGS

SPIRACLES

HEAD

ANTENNAE

9

CHAPTER 2

WHERE THEY LIVE
and What They Eat

Millipedes live all over the world. They like warm, damp places best. They often make rotting plants their homes. Some millipedes live under bark, stones, or leaves.

A millipede's world is dark. But that's OK. Millipedes don't need to see. Some don't even have eyes! They find their way using their antennae.

MILLIPEDE RANGE MAP

Millipedes live on every continent except Antarctica.

EATING DEAD LEAVES

EATING FUNGI

PUSHING THROUGH DIRT

Life Among the Dead

Most millipedes' homes are also their dinner. These creatures eat dead leaves, plants, and fungi. Millipedes munch their way through their hidden homes. Some push through the dirt like bulldozers.

Some millipedes eat small bugs and living plants.

TYPES OF MILLIPEDES

BARK DWELLERS

tiny bodies

live under bark

BORERS

have small, pointed heads

good at widening cracks in walls

BULLDOZERS

**long and thin
push through dirt with hard heads**

ROLLERS

**short bodies
roll into tight balls**

WEDGE TYPES

**not as rounded with plated bodies
use flat heads to open small spaces**

CHAPTER 3

FAMILY LIFE

In the spring, female millipedes make nests of dirt and their own waste. Then they lay up to 300 eggs.

Most mother millipedes don't watch over their eggs. After hatching, baby millipedes must take care of themselves.

Molting

Young millipedes have few legs.
But they soon shed their skin and
grow legs. This process is called
molting. Millipedes molt several times.
Each molt, they add more legs and
body parts.

**Millipedes often eat their
skin after molting.**

Millipede
LIFE CYCLE

Female millipedes
lay eggs.

Young become
adults in one to
five years.

Eggs hatch in a few weeks.

Young millipedes molt and grow more legs.

CHAPTER 4

THEIR ROLES
in the World

Millipedes are an important part of the food chain. Many animals eat them. Millipedes have several ways to protect themselves from these **predators**. Many curl up into balls to protect their heads and feet. One bristly millipede swipes its tail at attackers. The **bristles** attach to the predators. Most millipedes ooze poison from their **pores**. Some millipede poison is deadly to certain creatures.

Millipede Food Chain

This food chain shows what eats millipedes.
It also shows what millipedes eat.

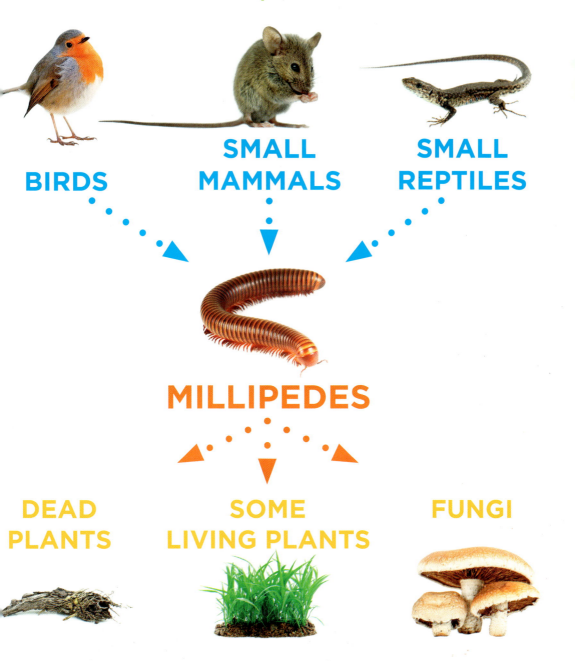

BIRDS

SMALL MAMMALS

SMALL REPTILES

MILLIPEDES

DEAD PLANTS

SOME LIVING PLANTS

FUNGI

A type of California millipede glows in the dark. Its glow tells predators to stay away.

Recyclers

Millipedes are good for the environment. They eat dead plants and animals. Their poop makes the soil rich. Moving through the dirt, millipedes make room for air and water. These actions help new plants grow. Millipedes are an important part of the world.

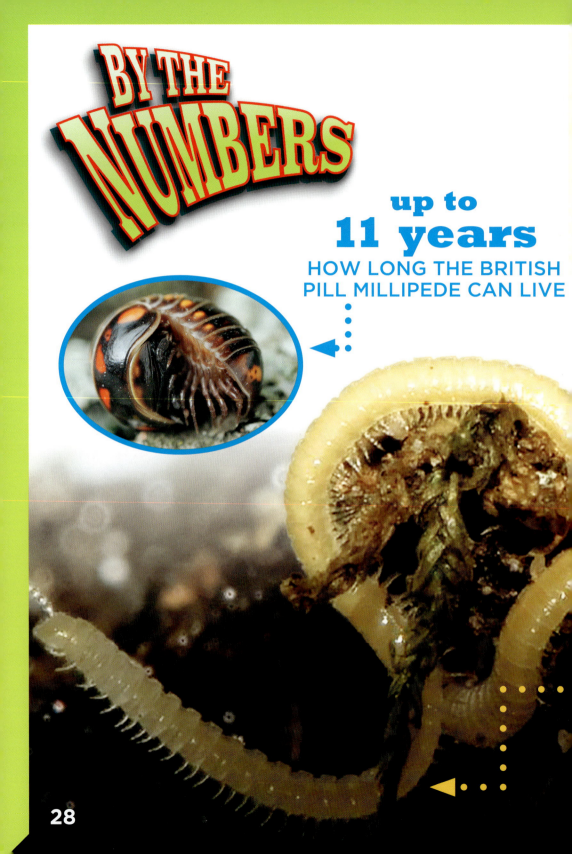

BY THE NUMBERS

up to
11 years
HOW LONG THE BRITISH
PILL MILLIPEDE CAN LIVE

about 7 feet (2 meters)
how long scientists think **prehistoric** millipedes were

200
NUMBER OF POISON PORES A CALIFORNIA CAVE MILLIPEDE HAS

HOW MANY LEGS AN ILLACME PLENIPES MILLIPEDE CAN HAVE

up to **750 legs**

GLOSSARY

arthropod (AHR-thruh-pod)—any of a large group of animals, such as crabs, insects, and spiders, with jointed limbs and a body made up of segments

bristle (BRIS-uhl)—a short, stiff hair

fungus (FUN-gus)—a living thing, similar to a plant that has no flowers, that lives on dead or decaying things

pore (POHR)—a tiny opening

predator (PRED-uh-tuhr)—an animal that eats other animals

prehistoric (pree-his-TOHR-ik)—existing in times before written history

segment (SEG-ment)—one of the parts into which something can be divided

spiracle (SPIR-uh-kuhl)—an opening on the body used for breathing

BOOKS

Rhodes, Wendell. *Millipedes.* Dig Deep! Bugs that Live Underground. New York: PowerKids Press, 2017.

Stewart, Amy. *Wicked Bugs: The Meanest, Deadliest, Grossest Bugs on Earth.* Chapel Hill, NC: Algonquin Young Readers, 2017.

Turner, Matt. *Tiny Creepy Crawlers.* Crazy Creepy Crawlers. Minneapolis: Hungry Tomato, 2017.

WEBSITES

Centipedes and Millipedes
easyscienceforkids.com/all-about-centipedes-and-millipedes/

Glowing Millipedes Revealed
kids.nationalgeographic.com/explore/nature/glowing-millipedes-revealed/

Millipede
a-z-animals.com/animals/millipede/

INDEX